KELP

The Health Giver

By the same Author:

THE BIOCHEMIC PRESCRIBER
BIOCHEMISTRY UP-TO-DATE
BUILDING A HEALTHY HEART
THE GROUP REMEDY PRESCRIBER
HEALTH FROM THE KITCHEN
HEALTH SECRETS OF ALL AGES
LADY BE BEAUTIFUL
THE NATURAL HOME PHYSICIAN
TRANQUILLIZATION WITH HARMLESS HERBS
ABOUT DANDELIONS

KELP

THE HEALTH GIVER

by

ERIC F. W. POWELL

Ph.D., N.D.

A book of HEALTH MASTER Knowledge

HEALTH SCIENCE PRESS

Wellingborough, Northamptonshire

This Edition (completely revised and reset) August 1968
Second Impression April 1969
Reset and reprinted January 1970
Second Impression September 1970
Third Impression October 1971
Fourth Impression October 1972
Fifth Impression June 1974

ISBN 0 85032 060 7

Made and Printed in Great Britain by
Weatherby Woolnough Ltd., Wellingborough
Northants, England

Contents

	Page
HEALTH FROM THE SEA	7
MEDICINAL VALUE OF KELP	8
KELP FOR NERVOUS DISORDERS	13
KELP FOR HEADACHES—INTERESTING CASES	14
KELP IN ARTERIAL DISORDERS AND HIGH BLOOD PRESSURE	16
INDIGESTION: THE DUODENUM AND PYLORUS	17
KELP FOR COLONIC WEAKNESS	19
KELP ACTIVATES THE IMPORTANT ORGANS	21
THE THYROID GLAND	29
FAT METABOLISM AND OBESITY	32
KELP TREATMENT FOR RHEUMATIC TROUBLES	33
KELP IN COUGHS, COLDS AND BRONCHITIS	34
HOW TO TAKE KELP	35
THE IMPORTANCE OF PURE FOOD	37
KELP AS A FIRST-CLASS FERTILIZER	39

Health from the Sea

Life on this planet has its genesis in the waters. "And God said, let the waters bring forth . . ." It seems logical to assume that as life began in the waters the seas must contain vital agents which promote all those biological activities which produce living forms, and that whatever comes from the seas should be rich in life-sustaining properties. The science of biochemistry proves that this is so, and of all the life forms found thriving in the seas the common seaweed known as Kelp deserves special attention.

Botanical description:

Fucus Vesiculosus.

Common names: Bladderwrack, Seaweed, Kelp, Bladder Fucus, Kelpware.

Appearance: Kelp grows among the rocks on most coasts. The perennial frond is coarse, yellow or brownish-green in colour and growing from two to three feet. It attaches itself to the rocks by branched, root-like, discoid extremities, developed from the base of the stalk. The frond is almost fan-shaped and like a strap at the base. In texture it is tough and leathery. Air vesicles develop in the substance of the frond and these "bubbles" may attain an inch in diameter.

Medicinal Value of Kelp

Kelp has been employed in medicine for a very long time by both the orthodox and the herbal schools. Later it found its way into Homœopathy. Recently, as a result of experiments, it has been found to possess an even wider range of therapeutical uses.

The Constituents of Kelp

The iodine content of Kelp varies according to the locality where it is gathered and also is affected by the season of the year. It is sometimes much higher than the figures quoted below. The content of other minerals also tends to vary slightly.

AVERAGE ANALYSIS

Moisture	8.00%
Protein (crude)	7.50%
Fibre (crude)	7.00%
Nitrogen (free extract)	42.20%
Fat (ether extract)	0.30%
Ash	35.00%

AVERAGE ASH ANALYSIS

Iodine	0.15-0.20%
Calcium	1.20%
Phosphorus	0.30%
Iron	0.10%
Sodium	3.14%
Potassium	0.63%
Magnesium	0.76%
Sulphur	0.93%
Chlorine	12.21%
Copper	0.0008%
Zinc	0.0003%
Manganese	0.0008%

Spectrographic analysis also disclosed the presence of traces of Barium, Boron, Chromium, Lithium, Nickel, Silver, Titanium, Vanadium, Aluminium, Strontium, Silicon.

Vitamins

Analysis discloses the following:

Riboflavin .. 1.5 Mg. per lb.	Niacin .. 25.0 Mg. per lb.
Choline .. 175 Mg. per lb.	Carotin .. 10 Mg. per lb.

Algenic acid is also present (Mannitol).

A study of Kelp analysis shows that it is a remarkable food containing more mineral matter and vitamins than many well-advertised proprietary health foods, with the additional advantage that these minerals are not added crude chemicals but "natural" in that they have been assimilated by the growing plant. Hence such minerals are far more likely to be acceptable to the human organism.

Students of Homœopathy and Biochemistry will be much impressed with what Kelp analysis reveals and realise the vast amount of good that must result from the daily intake of so many vital elements in a perfectly natural form.

It has also been stated that a small quantity of a sugar named Fucose exists in the dried Kelp. The air in the vesicles or bladders consists of a considerably higher percentage of oxygen and a lower percentage of nitrogen than is found in the atmosphere. It will be seen, therefore, why Kelp has such value as a medicine and fertiliser.

Plants are the only organisms that manufacture food, and the essential nutritional elements which we receive from animal flesh comes originally from plant life. As all the necessary raw materials are present in the oceans it will be seen why Kelp, so rich in these substances, is obviously such a vital food and remedy. The minerals it absorbs from the sea water and ocean bed in such abundance are presented to us in an organised colloidal state to be readily assimilated and utilised by the human body.

The old herbal school used Kelp mainly as an anti-fat remedy and for simple goitre, and until recently medical doctors employed it for the same purposes. We now know that seaweed contains far more elements of value than was originally thought. Many of the elusive trace elements and essential vitamins are also present, and these items greatly extend its range of usefulness.

The Homœopaths use this remedy for obesity, goitre, poor digestion, flatulence and obstinate constipation, and they advocate both material doses and the potentised preparation. Dr. Herbert Knapp, a homœopathic physician, came to the conclusion that he had found a specific for exophthalmic goitre in Fucus, and records cases of this disorder that he cured. His findings are supported and confirmed by a Dr. Foster, of Chicago, who states that he has never known it to fail when the patient was under thirty years of age. He adds that he never uses any other remedy for goitre. Herbalists and Homœopaths agree that the remedy must be taken for some time in order to produce ideal results, but as it is absolutely non-poisonous and is as much a food as it is a medicine, there is no limit to the length of time during which it may be taken.

I personally, have found goitre cases yield to the influence of Fucus, and I have also been able to reduce fat people to more normal proportions by the use of this remedy alone, plus a little common sense with dieting. The beauty of using Fucus for obesity is that, unlike the case with drugs, no harm is done. It is well to point out that if a thin person takes this remedy for a goitre or as a tonic or remedy for any other condition, he or she need have no fear that it will reduce weight still further. Fucus acts on obesity mainly through the thyroid which gland it tends to normalise. Thyroid trouble can cause obesity and it can also cause excessive thinness. Hence a normal thyroid helps to maintain normal weight. Indeed, I have known thin people put on weight when taking Kelp.

One function of the thyroid gland is to replenish energy which it accomplishes in co-operation with other endocrine glands. It has been found that there is a definite connection between the amount of energy available and our iodine intake. In Kelp we have a perfectly natural source of all the iodine we require.

Not only does Fucus act on the thyroid, but it most certainly has a remedial and normalising action on the sensory nerves, meninges, arteries, pylorus, colon, liver, gall bladder, pancreas, bile duct, kidneys, prostate gland, uterus, fat cells, testicles and ovaries. Few remedies have such a wide range of action and we are greatly indebted to the late Dr. Guyon Richards, the famous authority on

Radiesthesia, for discovering most of the new information relating to the action of Fucos on these glands and organs. However, it is well to point out that, while large doses can do no harm, Fucus will usually act better if taken in small doses, especially if any of the organs mentioned happen to be "diseased". In very chronic states the remedy may act better in the 30th homœopathic potency—five pills nightly until results are obtained; but I advise Fucus in this potency only under the guidance of an experienced Homœopath. For general purposes I suggest doses of the 1x or 2x potency in tablet form two or three times a day.

Kelp and the Tissue Salts

Fucus has an even better action in many instances and for general tonic purposes when it is blended with the tissue salt of Schuessler. The plant seems to act as a sort of vehicle for the cell salts and the results are most satisfactory.

For instance, I have known of cases where people have suffered from calcium deficiency. The use of calcium-rich foods and the tissue salt Calcium Phosphate produced no results; but when Kelp (we will call it Kelp from now on) was taken *with* the salt assimilation took place. I have also noticed similar effects when Kelp has been used with other indicated cell salts.

Kelp for Nervous Disorders

There is no doubt that some nervous complaints are due to a deficiency of certain cell salts, especially those of phosphorus, potassium, iron, sodium and the elements of lead, copper, silver and zinc. In other words, many nervous troubles are "deficiency diseases" and should respond to correction nutrition and a diet rich in the essential nerve building minerals. The nerve cases in which I have found Kelp to be helpful are too numerous to mention, but I recall one bad case of neurasthenia where the sufferer, a young man of about 24, was utterly exhausted. He had nervous dyspepsia with much flatulence, complained of pain in his back, was constipated and could not sleep except in fitful snatches. Kelp played a vital part in his total recovery.

Of course he was placed on a suitable diet, given gentle breathing and other exercises and supported with psychological aid; but there was no real change in his condition until he took Kelp with his meals three times daily. He also had the cell salts of Kali Phos., Natrum Phos., Natrum Mur., and Silica.

A lady of fifty who could not sleep because of "nerves" responded to Kelp after a few weeks, although I had great difficulty in getting her to take the remedy as she was an out and out naturist who regarded all herbal remedies as being "contrary to nature cure", forgetting that the cabbages, carrots and other vegetables in her garden were all wild herbs at one time.

An elderly gentleman with a nervous heart and very frightened about his condition responded to Kelp, but only after many weeks on the remedy. Experience suggests that the older you are the longer you have to take Kelp for remedial purposes. As I have said before, it is essentially a food and has no drug action as such, but helps to re-build the weakened organs.

One of the functions of iodine is to bring calmness to the mind and body by relieving nervous tension. When nervous tension is marked there is excitability and irritation, sleeping becomes difficult and there is a drain on the vitality. Owing partly to its iodine content Kelp reduces tension, produces relaxation and enables the system to store up vitality and reserve energy; concentration becomes easier owing to the more free flow of blood through the brain and there is less muddled thinking.

Kelp for Headaches—interesting cases

Recently I had a lady suffering from severe headaches. I found her nerves, digestive organs, kidneys and all other organs to be in a healthy condition. I also had to rule out psychological causes. She had good nights except when her headaches were very severe. At times she was almost frantic with the pain, which was located mainly at the base of the skull and often extended to the neck.

A test by means of Radiesthesia revealed trouble with the meninges. She then told me that as a child she had at one time suspected meningitis (inflammation of the meninges) which never developed; the doctor's drugs having "cured" the trouble in the early stages. Again using Radiesthesia I found that Kelp was the remedy. I gave this item in homœopathic potency for two weeks and the pain vanished. Similar results would, I believe, have taken place had she taken crude Kelp in small doses; but the cure would have taken longer to be achieved.

A young man came to me suffering from malnutrition and nightly headaches which, he said, nearly drove him crazy. The pains were accompanied by intense throbbing at the top of the head. I noted that he was very debilitated, thin, "nervy" and exhausted. I advised him to sprinkle half an eggspoonful of powdered Kelp over two of his meals, breakfast and lunch every day. Additionally I corrected his diet and managed to get him to practice deep breathing exercises and take a quick cold friction bath every morning.

Slowly the headaches subsided and he was soon sleeping well. An interesting feature of this case was that after being on Kelp for three weeks he began to put on weight, became less nervous and was conscious of more vitality. The former general organic sluggishness disappeared; he became more interested in his work and was a far more happy individual.

I must add that many cases of headaches when located in the base of the skull have responded to Kelp treatment. The same applies to those who suffer much distress from tension and pain in the neck. Neck pain and congestion may be due to the meninges or to trouble with the pneumogastric nerve. But whatever the cause it is rare to find a case in which Kelp will not be a help.

Press your fingers into the tissues at the base of the skull and at the sides of the cervical (neck) spine, and if it is painful you may rightly assume that congestion exists. This congestion may be causing nervousness, exhaustion, headaches or indigestion. In such cases the sufferer should take Kelp regularly for some time. Results are likely to be pleasing. I have reason to believe that Kelp has helped cases of migraine, but as other remedies were employed I have to say that the Kelp only played a part in removing the causes of the trouble.

Kelp in Arterial Disorders & High Blood Pressure

The increase in dangerous conditions due to high blood pressure has reached alarming proportions. There are certain cases where the heightened pressure may be due partly or wholly to nervous stress but by far the greater number can be attributed to degeneration and hardening of the arteries. The blood is the life and health depends on its purity and nutritional value plus its free and unobstructed flow throughout the entire organism.

As Kelp has a normalizing action upon the thyroid and para-thyroid glands we may expect that as a result of the thyroid being tranquillized when it is hyper-active the sufferer from high blood pressure will become less excitable, and that the possibilities of emotionalism adding to the arterial tension will be minimized. There is ample evidence that Kelp is a fine nervine and it is necessary when dealing with hypertension that the nervous system of the sufferer should be calm and the mind at peace. When tranquillization has been achieved it is much easier to deal with the blood pressure by any therapeutical measures.

Better function of the para-thyroid glands means that the system can take up and utilize mineral matter to the best advantage; in particular calcium, iodine and sodium, which all play a part in maintaining the health and elasticity of the arterial walls. An important point in favour of Kelp is that it does not interfere with the therapeutical activity of any other medicines. It is a true food-remedy and in no sense could be termed a drug. Kelp will often achieve good results when the much-advertised complex vitamin and mineral preparations seem to be totally ineffectual.

Kelp is an arterial cleansing agent and gives tone to the walls of the blood vessels. Hence it is helpful in some cases of arterial tension (high blood pressure). Practitioners believe that it helps to remove deposits from the walls of the arteries and restore their elasticity, thereby lengthening life. Those with low blood pressure need not be afraid to take Kelp, as by improving the general health the pressure will be normalised; and few agents can equal Kelp for general health-building purposes.

Indigestion: the Duodenum & Pylorus

On occasions we encounter cases of severe indigestion that will not respond to the usual treatment and remedies.

In such instances the pylorus is often at fault. Where there is contraction (spasm) or congestion in the pylorus, digestion in the stomach is held up and the prepared nutrients cannot pass as freely as they should from the stomach for the next stage of the digestive/assimilative process. Kelp is a most useful remedy for almost all troubles with the pylorus. We have to thank Radiesthesia for this discovery. May I repeat that for this purpose small doses act better than large quantities; indeed the small dose is superior in almost all instances and the pulverised herb is better than the extract.

Not long ago I had a severe case of dyspepsia which failed to respond to the usual treatment. I was convinced that the cause was in the stomach area and resolved to treat the pylorus. Kelp proved to be the remedy.

Another lady suffered from digestive trouble for many years. She suffered pain from even the smallest meal, vomited frequently and was a picture of malnutrition. I found that she had spasm of the pylorus as soon as food entered her stomach. After prolonged treatment and constant failure I tried Kelp. In due course she experienced gradual relief and is now in a fair state of health.

Some cases of indigestion are not due to stomach or pyloric disorders, but to duodenal ulceration or inflammation. When eating relieves pain for some time after a meal one may be sure that the duodenum is at fault. While I do not claim that Kelp will of itself deal satisfactorily with duodenal ulcers, I can say with certainty that this precious sea plant is a positive aid in such cases. Any remedy that tones the stomach, aids digestion and deals with excess stomach acidity is bound to have a healing effect upon the duodenum. As Kelp is an antacid it performs this service, and will be an additional aid to any other treatment.

Dr. John H. Clarke, the famous Homœopath, advises Kelp for indigestion, flushing of the face, full feeling in the stomach and headaches due to indigestion. There is hope for many chronic sufferers from indigestion if they will give Kelp a fair trial. I think I may safely say that at least some benefit will result, and possibly a total cure.

Kelp for Colonic Weakness

A host of physical disorders and even some forms of insanity are due to toxæmia coming from a foul colon. Poisons accumulate in this large bowel and are absorbed into the blood stream, causing debility, rheumatism, nervous disorders, kidney trouble and all those complaints due to disorganised, toxic blood. This truth so impressed that great surgeon, Sir William Arbuthnot Lane, that he devised a method of dispensing altogether with the colon. This is known as "Lane's Operation". However, Sir William realised that he was cutting out a necessary organ of the body, and that the operation did not remove the reason why the colon became toxic in the first place.

Reformed healers pay much attention to cleansing the colon with enemas and high colonic irrigation, and a great deal of good has been accomplished by such methods. However, these cleansing processes do not contribute very much to toning up the colon so that it can get rid of its accumulated poisons unaided. A toxic colon is a weak colon and needs to be built up.

There are many remedies that perform this service, but certainly not purgatives, which only weaken the organ still further. Among a number of very excellent colonic toning agents Kelp must take its place with the best. The reason why this sea plant has such a toning action is probably due to its high natural mineral salt content. The salts build the walls of the organ and the iodine, being highly antiseptic, deals with the toxic condition. For this reason Kelp is advised in all cases and diseases associated with auto-toxæmia (self-poisoning). All constipated people should take Kelp daily, and its action will be assisted if black Molasses is taken as well. Few people are really free from constipation and as health depends so much on clean, healthy, active bowels, it will be realised that everybody must benefit by the use of a food-remedy that acts in the manner of Kelp.

I find difficulty in citing special cases where Kelp has helped regenerate the colon, for I have given it to so many —scores require such treatment. A small book could be written dealing with this one thing alone, and I am confident that the human race could be helped along the road to regeneration by adopting a diet that would keep the colon clean and prevent auto-toxæmia. We are assured that the human body as a machine is perfect and that there is no law limiting life. But authorities agree that the one obvious cause for degeneration is the accumulation of toxic substances, having their origin in the colon, which obstruct and retard all the vital life processes and cause eventual disease and death. If as much attention was paid to internal cleanliness as to the washing of the exterior of the body, remarkable results would soon be obvious.

I will make mention of one case in which Kelp proved to be of immense value. A lady of fifty years had suffered for a considerable time from pain in the ascending colon. She was also constipated, was subject to sick headaches and always felt very weary. The painful colon kept her awake at night. Palpitation revealed the colon to be hard, knotty and prolapsed. Before placing her on my usual remedies I decided to see what Kelp would do if taken three times daily for two weeks. There was no noticeable effect during the first ten days, and then to my surprise, and to her delight, she reported that she was experiencing much less pain and was not so constipated. Her headaches were about the same, due no doubt to the accumulated toxic state. However, within a month the headaches had cleared and she required no other form of treatment. Such simple treatment; yet so effective!

Kelp Activates the Important Organs

Normalising the Liver

If you get up in the morning with a sick liver you probably have a fit of "the blues". On the other hand a morbid state of mind will tend to produce a sluggish liver. The liver is very sensitive to mental states, and the mind also reacts to the liver. To be bright and vital one must have an active liver. This organ is very important to normal health; much constipation, which is the cause of auto-toxæmia, is due to liver trouble.

Kelp is an "organ remedy" for the liver. That is to say it has an affinity for the organ and has a direct action upon it. The action of this remedy is to supply the liver with the salts it needs for normal function, and it also has a sweetening and cleansing effect. Note that Kelp acts on liver *and* colon, hence its value is obvious to all who are costive, toxic, depressed and "out of sorts". I have known many very obstinate liver conditions yield to the influence of Kelp, but of course the sufferer must feed sensibly if any good results are to be maintained.

The case of Mr. B. is interesting. He suffered from much indigestion and had occasional bilious turns. Almost every morning for some months he had risen from bed in the morning with a sick headache. For a time he had taken purgatives, salts and aspirins, which, although giving a measure of relief, failed to produce any lasting benefit; indeed the drug intake eventually made him worse. I put him on a fast followed by a cleansing diet. From this he derived much benefit, but continued to have indigestion and liver bouts at intervals. Eventually he took Kelp with certain cell salts at every meal and after some weeks his old troubles departed. In this case it will be noted that the Kelp not only acted on the liver, but also on the toxic colon, the gall bladder, kidneys and meninges. The latter usually come in when headaches are severe. I shall deal with the action of Kelp on the kidneys later.

Clearing the Gall Bladder

When speaking of the liver we naturally think of the gall bladder. As a rule anything which helps the liver aids the gall bladder, but not always. However, Kelp acts on both organs and has proved to be very effective in clearing an obstructed gall bladder. It is probable that the highly evolved sodium content of the remedy plays a large part in this connection. I cannot say that Kelp has any value in cases of gall stones as I have never tried it on its own. But it seems reasonable to suppose that any remedy which tones the liver and gall bladder will have some good effects in these cases and will at least help to prevent the formation of stone.

Aiding the Pancreas

People only think of the pancreas when they have diabetes. But many of those who suffer from indigestion and constitutional weakness may have a weak pancreas. I have already mentioned that some forms of dyspepsia can be traced to a faulty pylorus. Similarly, there are other cases where the fault lies in the pancreas. Whatever the reason for the indigestion it is good to know of a natural, harmless remedy which is likely to deal with all causes because of its action on the organs as a whole.

Kelp, taken regularly, will help to keep the pancreas

in order and may well prevent more serious organic disease from developing later on.

Very often in my professional work I encounter cases where, although there is no sugar in the urine, the pancreas functions poorly. Kelp is one of the best organ remedies for a sluggish pancreas.

I quote one case of a lady who was greatly debilitated. Her doctor had tested for diabetes, anæmia and all the debilitating disorders. Her endocrine glands appeared to be in order and she had no worries. Indeed one was surprised to find someone so bright who felt so feeble in body. My tests revealed a sluggish pancreas. Kelp with the addition of one other remedy effected a fairly rapid cure. She did not have to go on a difficult, restricted diet to obtain the desired results. Those who practise Radiesthesia place Kelp very high in the list of remedies for pancreatic troubles, and its constant use cannot possibly do any harm irrespective of age or condition.

The Bile Duct

We cannot leave our discussion of the action of Kelp on the digestive organs without giving brief mention of its value as a remedy for obstruction of the bile duct. But as this is of little interest to the layman, it will serve if I merely state that the action of this remedy is profound and that it has done excellent work in this connection to my certain knowledge.

Vitalizing the Kidneys

I find that the ancients paid rather more attention to

the health of the kidneys, than we do today. Of recent years we have found that the kidneys are not only eliminative organs; they also aid in assimilation and are partly responsible for adequate nutrition. Also, it must be remembered that the vital suprarenal glands are situated on the top of the kidneys. These glands secrete adrenalin, play a large part in supplying us with vitality, and meet the demands for extra energy at times of stress and exertion. Anything which helps to keep the kidneys healthy must, we assume, also aid the suprarenals.

Kelp certainly cleanses and tones these organs, and I have found it especially valuable when the kidneys are in an "irritable" state and when they are painful. Indeed, Kelp has helped me to clear up kidney cases that were very stubborn and that had failed to respond to other treatments—naturopathic, herbal and biochemic.

The use of Kelp for many of the troubles and organic weaknesses I mention will be entirely new to most readers, but I hasten to add that every statement has been proved by experienced practitioners, and even by broad-minded medical men who are interested in drug-free therapy. Because a remedy has been employed for certain purposes in the past it does not follow that we know all about its possibilities. We have to keep open minds and add to our knowledge through experience and new discoveries in the realms of diagnosis and treatment.

In common with all other practitioners, I have had some very serious kidney cases, and very stubborn some of them have proved to be. Sufferers from serious kidney disorders should consult a reliable practitioner and not rely on self-help only. However, minor ailments may be treated successfully at home provided the diet is sensible and Kelp is used as a remedy. It may be combined with other indicated herbs and will not antidote Homœopathic remedies (an important point).

I have in mind a lady who had suffered from weak kidneys ever since childhood. At the age of forty she was weak, listless and depressed, in constant pain and feeling that life was not worth living. She was restored to health in due course, and while Kelp was not the only remedy used it played a vital part in her recovery. She takes Kelp regularly to keep her kidneys in good order.

Toning the Prostate Gland

Kelp for prostate gland trouble? That surely is a new idea! The Herbalists and Homœopaths think of Sabal Serrulata, Thlaspi, Hydrangea, Stigmata Maydis, Erigeron and several other remedies—but not Fucus Vesiculosus. Yet this remedy will do much to normalize a weak and enlarged prostate gland. To the experienced biochemist the chemical composition of the remedy explains just why. Many men faced with operations for prostate trouble could be spared this ordeal if they took Kelp regularly, especially after the age of forty years.

The action of Kelp on the prostate is to improve the nutrition of the organ and the circulation of the blood through the tissues. "The blood is the life" and, in common with all other organs, the health of the prostate depends on the normal circulation of chemically balanced blood through its substance. Again I can say that the value of Kelp in this connection has been proved by the excellent results obtained; but it is necessary to take the food-remedy for some time and wise to keep it up.

A man in my district was saved a nasty operation by taking Kelp, and he was nearly seventy years of age when he came to me!

Help for the Uterus

The organ which develops into the prostate gland in the male becomes the uterus in the female; hence it is natural that we should find Kelp to be a most useful agent for toning up a weak uterus. Of the ten vegetable remedies I employ for weakness of this organ, Kelp is one of the best, especially when the sufferer is troubled with associated nervous disorders and depression.

In my opinion Kelp would be of considerable value to pregnant females and help to promote normal delivery. Its use would be likely to prevent abortion and by enriching the mother's blood with the salts required by the unborn, the latter is obviously going to be born a more healthy babe.

I have been able to help some ladies who have lost babies in childbirth and others who have not been able to carry for longer than about one-third to one-half the usual time. In one instance there had been three miscarriages. The next confinement was normal and she gave birth to a healthy child. In these cases, or I should say in most of them, Kelp has played an important part.

I also find that Kelp tends to normalize the periods in weakly women, and does much to banish the depression sometimes associated with the appearance of the menses.

The Testicles

Here again we find Kelp performing a service not formerly thought of. For hardening of the testes and simple impotency this remedy may be expected to produce good results if persisted with. It does not over-stimulate and cannot harm. No doubt the value lies in the improved local circulation of blood rich in cell salts.

I recall one case of a man with very painful testicles. Kelp removed the pain in a surprisingly short time. Cases of impotent men have been given Kelp treatment with varying degrees of satisfaction, but it must be remembered that the causes of impotency are complex and often include the psychological side; so too much must not be expected of Kelp or any other remedy in the treatment of this type of trouble.

The Ovaries

For ovarian disorders Kelp does as much as it accomplishes for weakness of the uterus. Quite often the two troubles go together and what aids one organ directly affects the others.

For ovarian pain, irregular menses, depression and even in some cases of anæmia, Kelp will be found most helpful. The food-remedy should be taken for several weeks or months, and the normalizing process, although slow, is in harmony with natural law and usually produces satisfactory results in due course. There is good reason to believe that the constant use of Kelp may prevent growths and ovarian cysts, but it is not held out as a cure for such conditions.

The use of Kelp for female depression alone is well worth a thorough trial. Life is made more interesting and all the family benefits when mother is bright and cheerful.

The Thyroid Gland

I have already explained the need for iodine in the treatment of thyroid gland trouble. The Greeks ate sponges and other sea plants to cure goitre, but it was not until 1849 that Chatin established a connection between iodine deficiency and goitre. Later, iodine was discovered in the thyroid and it was found that in people suffering from goitre there was an iodine deficiency. Goitre has been produced in animals by feeding them on foods lacking in iodine, and females fed on such iodine-free foods have produced offspring with goitres. By administering iodine the animals were cured.

With the exception of sea foods, most articles of diet are lacking in iodine. Fish oils contain iodine, but Kelp is probably the best source. Too much iodine may produce over-activity of the thyroid which leads to mental excitement and emotionalism, and this is one reason why I advocate small doses of iodine products. On the other hand I cannot say that I have ever noticed any hyperthyroidism from taking Kelp, and this may be due to the fact that the iodine in Kelp is only a part of a highly organised arrangement of salts. Goitre can also be caused by too much lime in the system. Indeed, any lack of chemical balance in the system can produce organic derangements, and for this purpose it is always wise to prescribe products prepared in Nature's own laboratory, or in that finely divided state prepared by the Schuessler Biochemists and Homœopaths.

Cretinism in children is due mainly to thyroid trouble. Hence such children should be given Kelp daily, and while they may never be normal the iodine rich food should produce some degree of improvement.

Thyroid Functions

It will be of interest to readers to give a list of the important functions performed by this vital endocrine gland. They are as follows:—

1. Secretes thyroxin.
2. Controls and regulates metabolism.
3. Vitalizes every cell of the body and enables them to respond to sympathetic stimulation.
4. Assists in the control of tissue differentiation.
5. Increases the power and rate of the heart function.
6. Controls coagulation time.
7. Increases urea and fluid secretion.
8. Stimulates and brightens the mind.
9. Controls and regulates the body fat.
10. Controls intestinal activity.
11. Aids the function of the pancreas.
12. Helps to harmonise the activity of the suprarenal glands.

13. Has a regulating influence on the ovaries and testicles.
14. Works in co-operation with the parathyroids, thereby regulating the action of mineral salts in the system, especially of calcium.
15. Acts in conjunction with the pituitary gland, thereby exerting a profound influence on metabolism in general. (The entire chain of the endocrine glands (spoken of as "the chain of life") works in vital co-operation, each influencing the others. The glands represent BALANCE in the body; therefore if balance is to be maintained on both the mental and physical planes of being, we must keep the glands normal.)

Hypofunction of the thyroid produces lassitude of mind and body, cretinism, slow growth in children, delayed maturity, obesity, female troubles, dry skin, dry lustreless hair and kidney weakness and disorders.

On the other hand, hyperfunction of the gland produces a completely opposite picture: over-sensitivity, mental alertness, emotionalism and over-activity everywhere in the organism. In extreme cases we get Graves' disease.

People who find that injuries and wounds heal very slowly need thyroid treatment with Kelp.

The endocrine glands are the chemical laboratories of the body and they perform their task properly when the necessary chemicals are supplied by the blood stream. As Kelp contains iodine and so many of the essential minerals it may be said to be a potent gland food.

Fat Metabolism & Obesity

As previously stated, the value of Kelp in obesity is probably mainly due to its action on the thyroid gland. But this is not wholly so. We find that this remedy has a normalising influence on the fat cells. So whatever the causes for obesity by using Kelp the two most important are being attended to. Keep in mind the fact that normal circulation of chemically balanced blood will do more than any other physical factor in keeping the entire organism in a healthy condition. Owing to its rich iodine and mineral salt content Kelp does as much as any single remedy can to produce and maintain a healthy stream of life.

Thin, undernourished people need have no fear that, as Kelp is useful in obesity, it will cause the scraggy person to lose more weight. Much the reverse. Practitioners have observed that while Kelp undoubtedly plays a part in fat reduction it also so encourages normal nutrition that the thin tend to put on more weight. I have records of successes in both obesity and malnutrition and Kelp has played an important role in producing a normal condition of body in all of them.

I recall the case of a very thin and nervous lady who was very much the odd one out in a family of jolly, stout people. The mother and sister wished to reduce their weight, while the painfully thin one wanted to be as buxom as the others. A little Kelp taken twice daily resulted in the stout ones losing weight and the thin one putting on several pounds of healthy tissue. She also became much more composed and contented.

Kelp Treatment for Rheumatic Troubles

I believe that more people come to me suffering from rheumatism, fibrositis and arthritis than from any other diseases. Rheumatism is not common to Britain (as is supposed, owing to our climate); it is wide-spread. I have travelled much and have always found rheumatic sufferers wherever I have been. The Australians certainly suffer from this painful disorder as much as we do in England; and even more from goitre!

Rheumatic ailments are due to an excess of certain acids in the system, and are usually associated with faulty kidney function plus a deficiency of the sodium salts. Orthodox medicine realises this and people are given massive doses of sodium compounds to counteract the acidity. Unfortunately cures are rare and as a result of this over-dosing of crude minerals digestion is ruined and kidney function disturbed; arterial disease may manifest and life be shortened.

The rational treatment for rheumatism consists in dieting sensibly and avoiding acid-forming foods such as white sugar, white bread, cake, pastry, many potatoes and "fancy" foods. Sufferers should partake of an abundance of the cleansing foods rich in mineral salts. I advocate plenty of fresh green salads and vegetables and also special foods which are exceptionally rich in certain salts. For example, Kelp and Molasses. Parsley and carrots are good for rheumatic sufferers, also celery and radish tops. Those who complain that they cannot take radishes will usually find that by eating the tops as well as the roots they will agree.

Rheumatic people should take an Epsom salt bath

two or three times weekly, and above all keep their bowels and kidneys in good order. Elimination of acids is the key to cure, and also the re-building of healthy body cells through the medium of correct foods and natural remedies. Cure cannot be expected unless the bowels function normally, and I have never found a case of constipation that did not respond to a breakfast of soaked prunes, a little "All-Bran", a good pinch of Kelp and a small teaspoonful of Molasses in a tumbler of hot water. Such a breakfast is not only a natural laxative, but it will supply the blood with the necessary antacids and cell salts essential to the removal of acidity, promote normal kidney function and bring vitality to the entire organism. Kelp should also be taken with every meal and the dieting persisted with. Like growth, all natural cures are slow, but they are sure.

Kelp in Coughs, Colds & Bronchitis

In spite of all the advances of medical science as yet no "cure" has been discovered for the common cold. It is true, however, that certain herbal remedies can make short work of this troublesome malady by promoting elimination through perspiration. Also the Homœopaths and Naturopaths can hasten cure of the acute condition and help to prevent colds by building up the vitality generally.

As with other ailments, those who suffer from continual colds, coughs and catarrh are suffering from a cell salt deficiency, and they also lack sufficient iodine.

I am not holding out Kelp as a cure for the acute cough and cold, but I am saying that the constant use of Kelp is very likely to so increase the resistance to these maladies that the system will in due course become free from them.

At the time of writing this small booklet I have many cases of coughs, severe colds and catarrh among my patients, and especially among the children. It is my expectation that by getting them to take Kelp and certain cell salts regularly they will all eventually become free from these vitality-destroying ailments which so far have defeated the best brains in medical science. Be it noted that a cold is an effort on the part of nature to throw off morbid encumbrances—a sort of systemic spring-cleaning process. Sane living and healthy exercise are important. Combine these things with the taking of a salt/iodine rich substance (Kelp) and the body will not have to make these repeated efforts at self-cleansing.

A Useful Ointment

Melt two ounces of "Vaseline" Petroleum Jelly and well stir in a teaspoonful of powdered Kelp. Mix again once or twice before the mixture has cooled. This preparation makes an excellent ointment for cuts, sores, boils and pimples. It is both antiseptic and healing.

How to take Kelp

Long experience has convinced me that most herbal remedies have a much deeper action when they are potentised Homœopathically. By this process the real healing power of the remedy is "liberated", and a tiny dose produces far better effects than the larger doses of the same item in crude form. I have employed both crude and potentised preparations for many years and am convinced that the potentised article has a much deeper and prolonged action on the system.

This is true of Kelp in potency. I have found that in cases where the crude article gave but limited aid, the employment of the 1x potency produced satisfactory results, especially when combined with indicated potentised mineral salts.

The system can deal with and take up a remedy in potentised form when it fails to react to the crude material dose. Also, when it comes to the matter of an unpleasant taste, potentising takes away much of this in the low potencies and eliminates taste entirely in the higher. Hence a potentised remedy is more pleasant to take, especially for children and invalids.

Medical Opinions

T. J. Lyle, M.A., M.D., Professor of Therapeutics and Materia Medica in Chicago Physio-Medical College, says of Kelp:—

> "This plant has a reputation as an antifat, claiming that it diminishes the fat without in any respect injuring the health. It influences the mucous membranes and the lymphatics. It is a gently stimulating and toning alterant.* It is one of those slow persistent agents that require time to accomplish the

* i.e., it alters the character of the blood and has a cleansing action.

desired results. It is stimulating to the absorbents and especially influences the fatty globules. Its best action is observed in individuals having a cold, torpid, clammy skin and loose flabby rolls of fat. It is an agent that gives better results in cases of morbid obesity than in those cases of a healthy character. It is best to begin with small doses—the urine becomes more abundant and the stomach is invigorated by its use.

"By its influence on the serous membranes it is valuable in cases of gout, rheumatism and dropsy. This agent seems to influence the starches and prevents their being formed into fats."

The above is an extract from Professor Lyle's famous text book: "Physio-Medical Therapeutics, Materia Medica and Pharmacy".

The late Dr. Guyon Richards mentions the value of Kelp in many conditions. Speaking of what he calls "reversed polarity" in the automatic nervous system he says that when such a condition exists it is "hell for the sufferer". For such condition he advises Fucus (Kelp). So when neurasthenics and other nerve sufferers are "going through hell" they are advised to take small doses of this humble weed from the sea. I must add that I can confirm Dr. Richards' discoveries from my own experience and observations.

The importance of Pure Food

It is positively alarming that, in these days of chemicalisation, it is almost impossible to obtain pure food. Our crops are poisoned with chemical fertilizers so that even the birds and insects are perishing. True, the chemicals destroy pests, but they also kill off those insects which are employed by nature for fertilisation. Crops and trees are also being poisoned with dangerous spraying, and many sincere practitioners have proof that some of the diseases affecting us today are a direct result of eating the vegetables and fruits from these "doctored" products of the earth. At long last the scientists and doctors are waking up to these very startling facts, and movements are abroad to try and put a stop to the rot. But a great deal of damage has been done, and it will take a very long time to restore the balance of nature. When possible, people should grow their own vegetables, and use home-made compost on their gardens or, order goods from farms where no chemicals are used on the soil. Such farms exist. The vegetables and fruits will cost a little more; but the reward will be better health.

Now we face yet another danger; the poisons from radiation due to atomic explosions and devices are slowly filling the atmosphere, and we have evidence that radiation is seriously affecting both plants and the animal creation. This is the price the human family has to pay for FEAR: fear of war, of enemies: and for the worship of material science. Doctors and politicians are literally dictated to by the enormous chemical firms who supply scientists with their destructive materials; doctors with their wonderful drugs, which create one disease whilst "curing" another; farmers with effective poisons for soil and crop treatment; housewives with dangerous detergents and cleansing materials.

I am all for the progress of true science and knowledge; but I am utterly opposed to science becoming a god. Let us get back to nature in matters that concern natural law.

About the only food unlikely to be contaminated by chemicals or dangerous radiation is that produced by the seas. So this is another good reason why, when we partake of Kelp, we are dealing with a pure, uncontaminated food-remedy. But for how long this freedom from contamination will last it is difficult to say, for there is no doubt that the waters will soon be affected by the radioactive material being dumped therein. However, the sea is a vast natural cleanser and purifier, and it will be some time before such a danger materialises.

It is worth noting that the doctoring of our water supplies with sodium fluoride and other chemicals is extremely dangerous. This is not merely the opinion of certain people. The dangers of fluoridation have been proved beyond all question in laboratory research and practical experimentation, and the long-term effects are quite alarming. Unfortunately authorities have to learn the hard way, and by the time they have been forced to use sound sense the harm done will be incalculable.

Kelp as a First-class Fertilizer

Kelp has been employed as a fertilizer for generations. It supplies quantities of sodium, potassium and other salts and also helps to oxygenate the soil. Poor soil may be regenerated within two or three years by liberal dressings of this seaweed. Indeed, if farmers would use more Kelp and far less chemical fertilizer the result would be better crops, free from disease and far more suitable for human consumption. This seaweed is also exceptionally rich in natural iodine, and for a very long time was the only source of the iodine used in medicine. Unfortunately iodine is now obtained in other ways and by "chemical processes". No doubt it is a better commercial proposition procured in this fashion, but the iodine thus obtained is not the same as that produced in Nature's own laboratory and has not the same biological effect on the human organism.

In some country districts Kelp is fed to animals with most satisfactory results. I am not sure if this is still the practice among farmers, but I know it was used in this manner when I was a boy, as I used to help gather quantities from the banks of the River Severn during holidays and make some pocket money by delivering my harvest to local farmers.

I find that adding Kelp to the compost heap greatly enriches the contents, making a little compost go a long way. It seems to act as a tonic and revives unfruitful trees; root crops improve in size and surface crops grow more abundantly.